Underwater Wonders

Ocean Animal Coloring Adventure

I0505231

This book belong to

_____

_____

Preface

Welcome to "Underwater Wonders: Ocean Animal Coloring Adventure"! In the pages of this coloring book, you will embark on a captivating journey into the mesmerizing depths of the ocean. Get ready to unleash your creativity and dive into a world teeming with fascinating marine life.

The vast and mysterious ocean is home to an astonishing array of creatures, from the smallest seahorses to the mighty whales that roam its expansive waters. With each turn of the page, you will encounter a new underwater wonder, waiting to be brought to life with the stroke of your coloring tools.

As you immerse yourself in the intricate illustrations, take a moment to appreciate the beauty and diversity of the ocean's inhabitants. Marvel at the vibrant colors of tropical fish, the graceful movements of dolphins, and the intricate patterns adorning seashells. Allow your imagination to soar as you envision the playful antics of sea otters, the majestic dance of jellyfish, and the hidden treasures lurking in sunken shipwrecks.

This coloring adventure is not only a delightful pastime but also an opportunity to learn about the incredible creatures that inhabit our oceans. Alongside each illustration, you will find interesting facts and tidbits about the animals, their habitats, and their unique adaptations. Let your coloring journey be a voyage of discovery, exploring the wonders of the deep blue.

Whether you are an avid coloring enthusiast, a nature lover, or simply seeking a moment of tranquility, "Underwater Wonders: Ocean Animal Coloring Adventure" offers a delightful escape into the enchanting world beneath the waves. Unleash your imagination, experiment with colors, and let your artistic expression flourish as you bring these magnificent creatures to life.

Remember, there are no boundaries to your creativity. Feel free to add your personal touch, embellish the scenes, and create a world uniquely your own. The pages of this book are your canvas, and the colors are your voice. So, grab your favorite coloring tools, find a cozy spot, and embark on an unforgettable coloring journey through the depths of the ocean.

Dive in, explore, and let the "Underwater Wonders: Ocean Animal Coloring Adventure" begin!

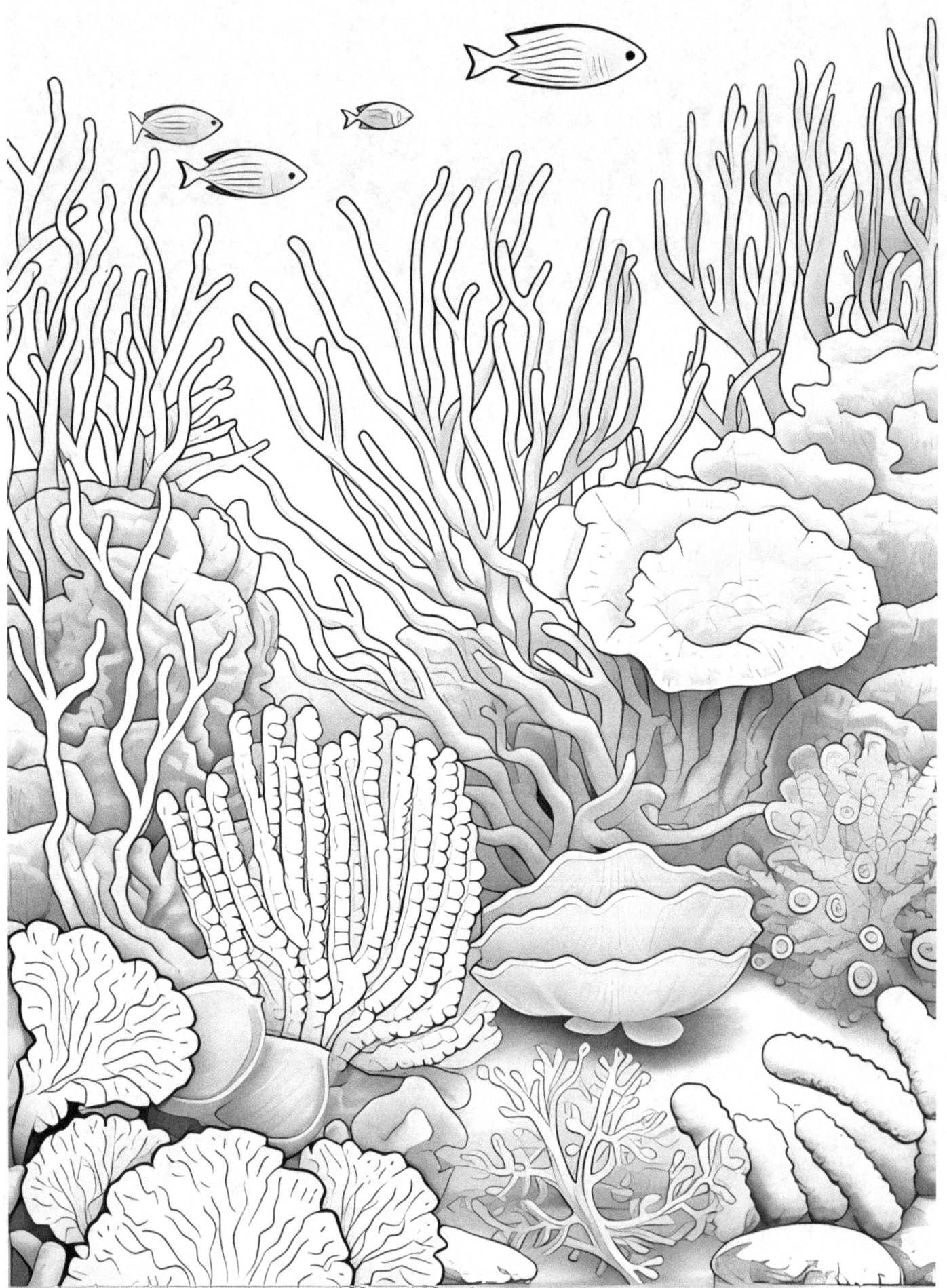

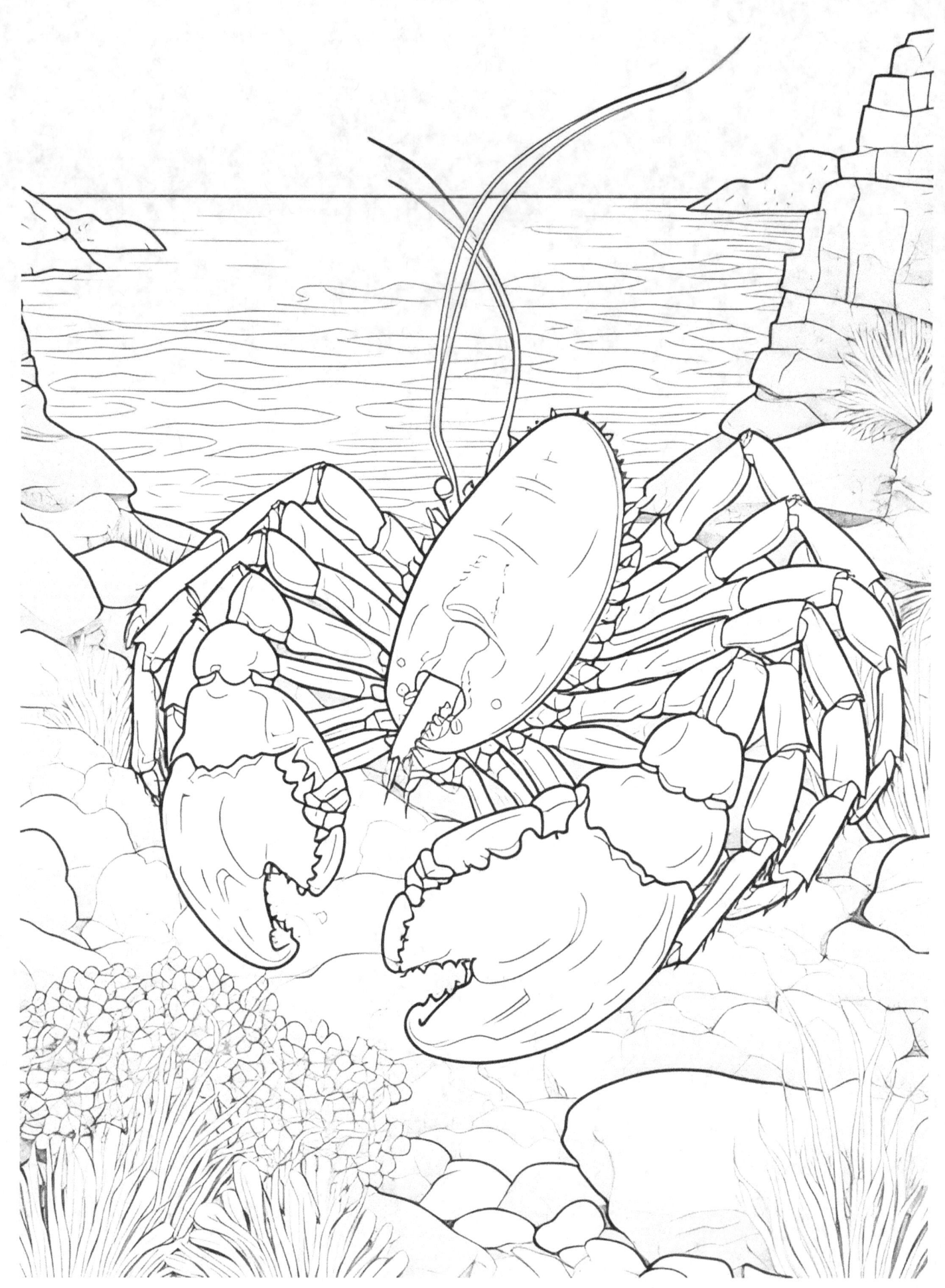